Spot the Difference

Ears

Daniel Nunn

Heinemann Library
Chicago, Illinois

Photo research by Erica Newbery
Designed by Jo Hinton-Malivoire
Printed and bound in China by South China Printing Company
10 09 08 07 06
10 9 8 7 6 5 4 3 2 1

Library of Congress Cataloging-in-Publication Data
Nunn, Daniel.
  Ears / Daniel Nunn.
     p. cm. — (Spot the difference)
  Includes bibliographical references and index.
  ISBN 1-4034-8473-2 (hc) — ISBN 1-4034-8478-3 (pb)
  1. Ear—Juvenile literature. I. Title. II. Series.
  QL948.N86 2007
  591.4'4—dc22
                              2006007238

Acknowledgments
The author and publisher are grateful to the following for permission to reproduce copyright material:
Ardea pp. **10** (John Daniels), **11** (Ken Lucas), **12** (Bob Gibbons), **14** (blickwinkel), **15** (Adams Picture Library), **20** (bilderlounge); FLPA p. **18** (Minden Pictures/ZSSD); Getty Images pp. **4** (Stone/Time Flachs), **21** (GettyImages/Photodisc Red/PNC); Nature Picture Library pp. **5** (T.J. Rich), **6** (Mike Wilkes), **7** (Lynn M. Stone), **8** (Anup Shah), **9** (Meul/ARCO), **13** (David Pike), **16** (David Kjaer), **17** (Bruce Davidson), **19** (Dagmar G. Wolf).

Cover image of a bat-eared fox reproduced with permission of Nature Picture Library/Graham Hatherley.

Contents

What Are Ears?

ear

Many animals have ears.

Animals use their ears to hear.

Where Are Ears Found?

Most animals have ears on their head.

This is a donkey.
It has ears on top
of its head.

This is a chimpanzee.
It has ears on the sides of its head.

ear

This is a cricket.
It has ears on its legs.

Different Shapes and Sizes

Ears come in many shapes.
Ears come in many sizes.

This is a fox.
It has big ears.

This is a sea lion.
It has small ears.

This is a wolf.
It has pointy ears.

This is a koala.
It has round ears.

This is a rabbit.
It has floppy ears.

Amazing Ears

This is an owl.
It can hear from high up in the sky.

This is an elephant.
It can flap its ears to keep cool.

This is a hippo.
It closes its ears to keep water out.

This is a bat.
It uses its ears to find its way.

Ears and You

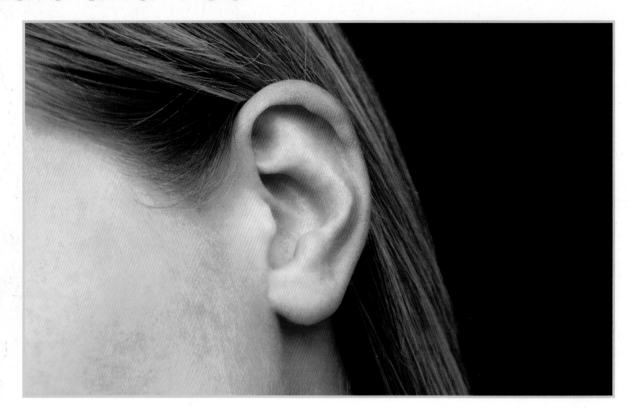

People have ears, too.

People use their ears to hear.
People are like other animals.

Spot the Difference!

How many differences can you see?

Picture Glossary

 **floppy** soft, not firm

 **hear** listen to sounds using your ears

 **pointy** sharp at one end

 **round** shaped like a circle

Index

Note to Parents and Teachers

National science standards recommend that young children understand that animals have different parts that serve distinct functions. In *Ears*, children are introduced to ears and how they are used to hear. The text and photographs allow children to recognize and compare how ears can be alike and different across a diverse group of animals, including humans.

The text has been carefully chosen with the advice of a literacy expert to enable beginning readers' success while reading independently or with moderate support. An animal expert was consulted to provide both interesting and accurate content.

You can support children's nonfiction literacy skills by helping them to use the table of contents, headings, picture glossary, and index.